CONTENTS

Distributed in the UNITED STATES to the Pet Trade by T.F.H. Publications, Inc., 1 TFH Plaza, Neptune City, NJ 07753; on the Internet at www.tfh.com; in CANADA by Rolf C. Hagen Inc., 3225 Sartelon St., Montreal, Quebec H4R 1E8; Pet Trade by H & L Pet Supplies Inc., 27 Kingston Crescent, Kitchener, Ontario N2B 2T6; in ENGLAND by T.F.H. Publications, PO Box 74, Havant PO9 5TT; in AUSTRALIA AND THE SOUTH PACIFIC by T.F.H. (Australia), Pty. Ltd., Box 149, Brookvale 2100 N.S.W., Australia; in NEW ZEALAND by Brooklands Aquarium Ltd., 5 McGiven Drive, New Plymouth, RD1 New Zealand; in SOUTH AFRICA by Rolf C. Hagen S.A. (PTY.) LTD., P.O. Box 201199, Durban North 4016, South Africa; in JAPAN by T.F.H. Publications, Japan—Jiro Tsuda, 10-12-3 Ohjidai, Sakura, Chiba 285, Japan. Published by T.F.H. Publications, Inc.

MANUFACTURED IN THE
UNITED STATES OF AMERICA
BY T.F.H. PUBLICATIONS, INC.

THE OSCAR'S APPEAL

GOOD POINTS AND BAD POINTS

Although it can be argued, endlessly and without resolution, just which fish is the most beautiful in the world, we must confess that the oscar is not in the running for that honor. (If you already own an oscar, it is recommended that you not let him or her read these lines, as the species is known to be both prideful and sensitive!) Oscars may have a certain mystique in their appearance, and many people may regard them as handsome; however, most assuredly no sane and sighted person would nominate an oscar for an Oscar in the "most beautiful fish" category. In spite of this fact, oscars have consistently maintained a very considerable popularity among tropical fish enthusiasts.

Oscars of today have been developed into many color varieties like this red one. Their personality has not changed, however. Photo by Dr. Herbert R. Axelrod.

This popularity is all the more remarkable when we take into account some of the disadvantages of keeping one of these fish. The main one, of course, is that oscars attain a length of about a foot, or perhaps a little more, in the home aquarium. And their growth is fast. If the young are fed heavily, you can almost see them grow! Size means that tank maintenance is going to be more difficult. Not only will you need good filtration, you will also need to make regular and copious water changes.

Although oscars are not as fiery and combative as most other cichlids, they prey on fish. For that reason, if you start an aquarium as a community tank of the usual assortment of tropical fish and include an oscar, you will soon have a tank of just one fish. The ironic thing about this fact is that some of the oscar's former tankmates may very well have picked on him at one time until he got big enough to eat them! An interesting point, though, is that oscars are not belligerent. They are just predatory. The other fish are their "cherry pies," and there is nothing "personal" about the fact that they consume them!

While not beautiful, baby oscars are as cute as puppies, and dealers must be conscientious about letting potential buyers know what they are getting themselves into, as a baby oscar in the typical community tank will very soon look upon all of his tank-mates as *hors d'oeuvres*. With my very first oscar, I had the best of intentions to leave him in my community tank only long enough to let him get just a tiny bit bigger in size, and then I would move him. I knew that I had procrastinated too long when I saw him swimming with a tetra sideways in his mouth, like a dog carrying a bone!

Keeping in mind the drawbacks of keeping an oscar, which primarily consist of their large size and predation upon other fish and the fact that they are not what one could really call beautiful, a question arises. Just what is the appeal of an oscar? Why have they been so popular for so long? As a long-term oscar lover, I am going to attempt, however inadequately, to answer that question.

First, of course, there is the famous oscar personality. These fishes are not only unusually intelligent (much like a parrot—but without the noise!), but they have a particular appeal. The young are like puppies, but so are the adults. They have a rounded appearance, even as adults, that makes them almost cuddly in appearance. An oscar kept as a pet very quickly learns to recognize its owner. It will "beg" for food from its keeper while eyeing strangers with proper suspicion. In fact, this food drive and propensity to beg for food can actually be utilized, for owners so inclined, to teach the fish tricks. I have seen oscars that would roll over for food. And I remember one that would leap out of the water like Shamu to take food. Usually oscars are "one-trick dogs," and it takes patience to train them, but the result is nonetheless remarkable.

Just how pet-like an oscar becomes is somewhat dependent upon the individual oscar, for there are individual differences within the species. And it is partly dependent upon the attention given the animal. We have known of many animals which actually

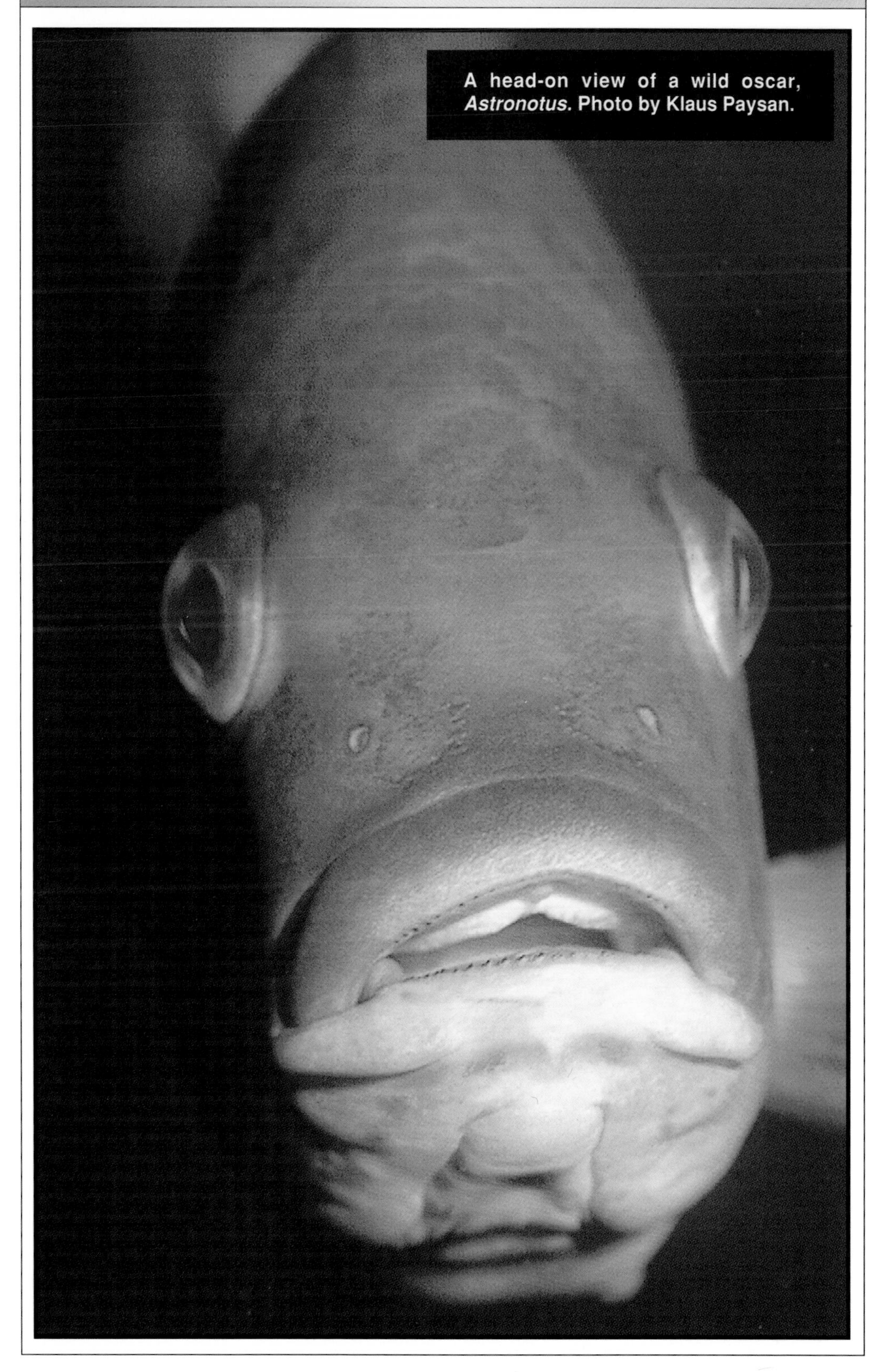

A head-on view of a wild oscar, *Astronotus.* Photo by Klaus Paysan.

allowed themselves to be petted by the owner—but they would bite at the hand of a stranger who tried the same thing.

Obviously, part of the oscar's personality is its intelligence. It is difficult to assess a fish's intelligence, as no valid intelligence test has yet been developed for them. (And some would argue the same for humans!) Even so, many people in a position to know consider the oscar the Einstein of fish. Certainly, the family to which the oscars belong, the Cichlidae, is composed of many fishes that many persons, including ichthyologists, consider as perhaps the most intelligent of all fish species.

Besides personality and intelligence, oscars have an appealing appearance, even if they are not in the beautiful category. First, there is the roundness of course, which we will discuss more at a later time, and it has a very definite appeal to it. There is also the quality of the scales. They are quite minute, and this gives the oscar a velvety appearance. In fact, an alternative popular name for the oscar for many years was the "velvet cichlid."

Finally, there is a variety to be had in oscars. That is, this is a species that has been in demand for many decades. For that reason, the fish have been cultured in aquaria and fish farms for many generations. Thus, artificial color forms and varieties have been developed. There are some old fuddy duddies (such as myself!) who are not too happy about artificial varieties, but the point is that the oscar is available in many forms. There is,

A trio of albino oscars. Note that they all have different color markings.

The most attractive oscar is the red tiger oscar which was probably developed in Thailand.

for example, the red oscar, the tiger oscar, and a long-finned variety of each. The future will probably bring even more varieties.

Thus, if you don't like a fish that matures to a velvety green, black, and red coloration, you can choose other options. Again, I will have more to say about the evolution of the oscar's natural form and colors later.

Finally, there is the matter of size. I know, I have already listed that among the disadvantages. However, many people find a special charm in big fish. They can easily be seen from across the room, and the delicate features of the fish can be more readily seen, too. And, of course, there is nothing that attracts the attention of visitors like a large fish. Guests may often ignore or only give scant attention to a beautiful community aquarium. But they are highly unlikely to ignore a large oscar. In fact, you can bet that they will want to know all about the fish.

Of course, you don't have to keep just one oscar. You can keep several and eventually end up with a mated pair. While you might not get the personal attention from a pair of fish that you would from an individual, you will find the oscar's family life nothing short of fascinating. The cichlids are famous among both hobbyists and ichthyologists for their parental care, and oscars exhibit this trait in its full magnificence.

The pair will rearrange the tank and clean off many rocks or spots on the glass before finally deciding where they are going to

The speckled red with black fins is a wonderful new color variety which has not gained favor and has probably been left out of the range of new oscar color varieties.

place their adhesive eggs. Both parents tend the eggs, fanning them to keep debris off and to provide a circulation of oxygen to them. They pick out any eggs that are going bad and continue to tend the rest. They take turns guarding. When the young hatch out, they place them in a large pit that they have already prepared for them. During the "wiggler" stage, when the young are unable to swim, the parents continually transfer them from one pit to another. The purpose behind all of this transferring is probably to give the young ones their "daily bath," as the parents cleanse the young of any debris and fungus as they mouth them during the transfers between pits (or "nurseries," as some hobbyists call them).

When the young hatch out, the parents herd them around to find food. They even chew up the food that you feed the adults and spit it out for their babies. Such tender care is rare enough among mammals, let alone fish. Best of all, the young will be readily purchased, as dealers are well aware that there is always a demand for oscars.

Oscars are easily spawned in empty aquaria providing the bottom is slate. This wild-type oscar is guarding his spawn. Photo by Joe Kislin.

Whether kept singly or as a pair, oscars have definite ideas about their environment. My wife and I changed the background of

The purple oscars developed in Europe from a sport of the red, never was fixed and is not available at this time. Photo by Dr. D. Terver.

the large tank in which we housed a mated pair of oscars, and both of them huddled toward the front in a mute but eloquent protest of our meddling. I remember seeing one oscar kept singly who claimed an area outside his tank as his territory. This consisted of a bulletin board on the right of the tank and a bird cage of budgies on the left. The fish acted for all the world as though this was part of his domain, but you would have to see it to appreciate it.

Young oscars which are not showing their characteristic eye-spot as yet. Photo by Burkhard Kahl.

As you can see, oscars do have a few things going for them. In fact, the very name that we utilize for them says something about their personality. The actual naming of this species with a popular name is lost to aquarium history. Perhaps it was just a misunderstanding of the scientific name, *Astronotus ocellatus*. But it could just as well have been that some hobbyist liked his personal piscatorial pet enough that he gave it a name, such as "Oscar." Eventually, it became the popular name for all oscars. Of course, we don't know that such a scenario is true, and we never will be able to to verify it. But it is as good a story as any.

The name itself bespeaks the very deep affection which this fish has always aroused. It is unthinkable that it would be known by any other. The oscar is truly in a class by itself. It is no wonder that tropical fish hobbyists from all generations have held it in such high esteem. When kept by itself as a pet, the oscar truly fulfills that capacity. They become as much a part of the family as any cat or dog. In fact, I have seen oscar pets that were completely spoiled, moody, demanding, and tyrannical. In short, they seemed totally convinced that they were part of the family!

Large Brazilian birds like the jabiru, *Jabiru mycteria*, feast on fishes. They do well on catfishes and oscars and other slow-moving fishes. Photo by Carlos Ravazzani and his colleagues from their lovely book *PANTANAL BRAZILIAN WILDLIFE*.

OSCARS IN THE WILD

LIFE IN THE AMAZON

Actually, oscars are found throughout the Amazon and Orinoco drainage systems in South America, including many named tributaries. During the rainy season these two drainage systems have a water connection, so there is some exchange of fish fauna during this time. Evidence of the success of the oscar is that it is a widely dispersed fish species throughout northern South America. The point of this chapter is to portray what life is like in the wild for this species and why it has developed some of its features and characteristics.

A one-word description for life in these ancient water-ways is "competitive!" There has been plenty of time for all manner of species of fishes and other animals to evolve. Thus, it is estimated by ichthyologists that there are more species of fishes in South America than in the entire Atlantic Ocean. (That is not to say that there are not more fish in the Atlantic Ocean in terms of the entire biomass.) There are, therefore, several species of piranha, as well as giant catfish, not to mention caiman (the South American version of the alligator) with which the oscar competes as a predator.

During the rainy season, rivers in South America expand to such a degree that the forests are flooded to the tree tops. This situation greatly increases the foraging area of the fish species, and many of them spawn at this time. When the dry season begins and the rivers contract, areas of the former "lake" become isolated, and the competition among the fishes is intensified as the fish become more crowded together. The bird predators have a "field day" (which actually consists of many weeks). The piranhas feast on the other fish as they become stranded and distressed. Then as the waters recede even further, the caimans feast on the piranhas.

The only fish to survive this receding of the waters are certain

You can look at bodies of water and know that cichlids can exist in them. This is a cichlid-supporting river in Madagascar. Photo by Dr. Melanie L.J.Stiassny.

catfishes (which are able to dig sustainable mudholes in the substrate and banks of the temporary water-way) and the fish which are intelligent enough to find their way back to the main waters before it is too late. Need I say more? Few oscars are left stranded in these doomed ponds.

Still, the oscar must make its living competing with such fierce predators as piranhas, not to mention other cichlids that are quite tough in their own right, such as the red terror (*Herichthys festae*) and the green terror (*Herichthys rivulatus*). Scientists have only recently done research that throws light on how the oscars succeed in doing just that.

Oscars are predators on small fish, which makes sense. Large fish are impossible to swallow, so the oscars take no interest in them. The small fish consist of the young of large species or adults of species which are smaller. Of course, such individuals have themselves evolved to escape predation as much as possible. Predators must be fast or clever in order to capture them. Oscars have evolved to fill the latter niche. They explore areas in which such small fish might hide, and such behavior can be seen in the aquarium. It all adds up to a

In the Amazon River system are thousands of streams and lakes that drop as much as 30 feet during the dry season, thus removing the protection from many fishes and subjecting them to the appetites of larger fishes, piranhas, caimans and to men who use nets. This is a typical locale on the Amazon. Photo by Dr. Herbert R. Axelrod.

This pond is drying out and is filled with fishes which have drained from the rain forest. Oscars found in this pond were well fed. The smaller fishes were gobbled up by the oscars, but the oscars eventually were eaten by caimans which could move from pond to pond. Photo by Dr. Herbert R. Axelrod constant curiosity. That is, it is a behavior which seems curious to us, and after all, we can only interpret behavior in terms of ourselves. Certainly, it is true that fully–fed oscars will still check out little nooks and crannies throughout the aquarium. They look most curious while doing such things, and who is to say they are otherwise?

Another method the oscar uses for finding a meal is to hang near the surface of the water where many of the smaller fish swim. Often this area is covered with floating plants or other debris. The rounded form of the oscar (and the overlapping of the dorsal, caudal, and anal fins to facilitate the rounded appearance) is believed to have evolved to blend in with such surroundings. The markings on the wild oscar, with its irregular blotches of greens, black, and red, also serve as a type of camouflage to disguise the fish from its prey. But oscars have another problem. In a word, it is piranhas.

Piranhas are a unique and specialized type of predator. Current thinking is that they evolved from fin nippers. That is, they evolved from fishes that supplemented their diet by nipping the fins of fish and swimming quickly away before experiencing retaliation from the fish whose fins had been nipped.

The rain forest in Brazil which is the habitat of the oscar on the Rio Negro. This is high water time. Oscars can be caught with hook and line during this period of heavy rains. Photo by Dr. Labbish Chao.

Fishermen in the Rio Negro habitat of oscars are able to collect hundreds of them per day using large nets and chasing the oscars into the nets. Photo by Dr. Herbert R. Axelrod.

Even today, the fiercest species of piranhas is most likely to feed by taking a big chunk out of a big fish which is preoccupied with something else. An oscar lurking near the surface of the water, even if it is partially disguised by plants and debris, is a sitting duck for piranhas. This is especially so when the tail is hanging ever so temptingly down.

After many weeks of observing oscars in their natural habitat, scientists have concluded that oscars have partially developed the rounded appearance in order to make the tail not quite so obvious. In fact, with the eye spot or ocellus (from which this species gets its specific name *Astronotus ocellatus*) in the tail or posterior part of the dorsal fin together with the rounded rear form, oscars actually appear as though they are head down from the surface of the water when they are head up. Scientists noted that the piranhas were not inclined to nip the fins of the oscars. They were apparently fooled by the clever camouflage evolved by oscars and possibly inhibited by the eye spot as many predators are. (Eye spots are known to inhibit predation, and they are utilized among a variety of animals to that end. One of the most fascinating examples is the caterpillar which has an eye spot in one of its folds which it displays when a bird approaches.

This aerial view shows the Rio Negro immediately after the rainy season in December, 1996. There are literally thousands of ponds in the area and these ponds are the favorite habitat of the Rio Negro variety of oscar. Photo by Dr. Herbert R. Axelrod.

This very successful defense has been observed to stop a would-be predator in mid-flight, causing the bird to veer away. Apparently, many animals are "hardwired" with a fear of large eyes, because large eyes usually mean a large animal, and, in many cases, a predatory one.) Some oscars have eye spots in the dorsal fin, and we will return to a short discussion of that subject.

Inside the Amazonian rainforest there are small intermittent ponds (ponds that dry out once in a while). These are usually laden with oscars. Photo by Dr. Herbert R. Axelrod.

Knowing the apparent evolutionary pressures which caused the oscar to develop the roundness, the color, and the eyespot that we all like should not diminish our appreciation for them. To the contrary, such knowledge enriches our enjoyment of the fish.

Part of the oscar's appeal is that they are members of the cichlid family (Cichlidae), for all members of this family tend to be a little more intelligent than other fish, and they care for their young in some way. Cichlids have had a "tough row" to hoe in the South American drainage systems, as these were very old and established freshwater regions in which primary freshwater fish had taken nearly all the niches available. You see, cichlids are secondary freshwater fishes. That means that they have evolved from a marine ancestral form that slowly adapted to a freshwater environment and subsequently evolved into many other species. That cichlids have succeeded in such a difficult situation is a testament to the hardiness and adaptability of the family.

Their capability of dominating a habitat has been demonstrated when they have been the pioneer fish, as in the Central American areas, which are relatively new freshwater habitats. (The Central American isthmus has only "recently" been elevated, speaking in geologic terms.) Since cichlids have evolved from an ocean ancestral form, many species were able to migrate across short expanses of the ocean and, thus, be among the first species of fish to colonize the newly-opened freshwater areas of Central America. They evolved to fill nearly every niche, and there are more species of cichlids in Central America than of any other family.

For the avid oscar lover, there are tropical fish collecting boats available for rental in Manaus. This boat is called *Cichla ocellaris*, a favorite game fish in the Amazon (also a cichlid like oscars). It costs about $500 per day for rental which includes food, room and guides. Photo by Dr. Herbert R. Axelrod.

Similarly, cichlids have been more successful in the Great Lakes of Africa, and the scientists do not know with certainty just why that is. Among the reasons postulated has been the presence of parental care of the eggs and fry by the parents. (That way it is not essential for the fish to be dependent upon flowing water to keep the eggs clean and safe, as in the upper reaches of a stream, an important advantage in colonizing expansive or deep lakes.) The intelligence of cichlids has also been put forth as an advantage that they have had over other freshwater fishes in colonizing new areas.

As cichlids, oscars come by their intelligence naturally. But the trouble with many cichlids is that they have an extremely bellicose disposition. That has evolved with a purpose, of course, as the individuals which were particularly aggressive were able to get the most desirable spawning areas.

Cichlids vary as to how the young are cared for. Some lay adhesive eggs on a hard surface in the substrate and guard them, just as oscars do. Others, including even large cichlids, dig

Lake Malawi in Africa is famous for its hundreds of different cichlid species. It would have been interesting to find out how oscars would have evolved under these circumstances. Photo by Dr. Herbert R. Axelrod.

an underground system of tunnels, and it is there that the eggs and the eventual young are guarded. Others spawn in caves formed by rocks. Several species use the leaves that have fallen in the water off fig trees to spawn upon. They thus have a mobile egg platform that they can move around as danger threatens. Many other cichlids are mouthbrooders. That is, they take the eggs in the mouth and keep them protected there, all the while tumbling the eggs to clean and aerate them as they are being incubated. In most mouthbrooding species the female incubates the eggs, but in some the male does the job, and in some both parents incubate the eggs. There are even some species in which the parents are partial mouthbrooders, in that they tend the eggs and fan them and only later do they take the young into their mouths for protection. In some of the African lakes, mouthbrooding cichlids incubate the eggs and fry for so long that the young are released fully able to swim, looking very much like a miniature of the adult. You get the idea. Parental protection among the cichlids has evolved in a multitude of directions.

Fortunately, oscars are not as aggressive as many of the other cichlids, and this is undoubtedly one of the major reasons why it has maintained such popularity among the general public.

An underwater view of Lake Malawi, Africa shows cichlids diving in and out of rocky crevices. Cichlids love the security of rocky caves. Photo by Dr. Herbert R. Axelrod.

Lake Malawi, Africa. The stones and vegetation are often under water during the rainy season. The rocks continue into the huge depths of this cichlid lake. Photo by Dr. Herbert R. Axelrod.

Nevertheless, oscars have to raise their young among all the predatory animals in South America about which we have made some mention. One of the ways that they accomplish this is with close parental cooperation. Both parents care equally for the eggs and young, taking turns to relieve each other when fanning the eggs, for example. With many other cichlids, only the female guards the eggs, while the male guards the outer perimeter. With oscars, the extreme predatory pressures on the eggs has forced the parents to stay quite close to them and to coordinate their defense with precision.

Possibly because of the need for close cooperation, oscars may stay together as pairs, even in the wild. This is not known with certainty, as some cichlids only pair during the breeding process. But in many cichlids, the male is substantially larger than the female. The scientific thinking on that is that the male needs to be larger, as his main duty is to provide the outer protection against marauding fishes. In the case of oscars, the female is as large as the male, and the two are indistinguishable. That is, it is difficult, if not impossible, to tell females from males just by looking at them. Fortunately, the fish can tell the difference, and they very likely remain spawning partners for life. Even if they don't, they establish a close bond and form an effective defense for the eggs and the young. When confronted with myriads of tiny fish trying to find an opening to dash in and grab a bit of the eggs, the parents will take a head-to-tail position and nearly completely cover the eggs. Larger fish are fiercely driven off. And yes, oscars even chase piranhas away from their progeny!

Whatever the evolutionary reasons, oscars are open spawners. That means that there is no attempt to dig a cave or to find one. A large pair of oscars may lay up to 3000 eggs on exposed stones or even large pieces of wood, and that is an indication of the predatory pressure on them. Even with the very capable defense provided by the parents, not many of those eggs are going to make it to adults in the wild. Of course, a lot of the young are going to be lost after they have dispersed from the auspices of parental care. The markings on young oscars is obviously camouflage, as all of the lines serve to break up the definite outline of the body. It is believed that oscars tend their young for about a month in the wild. By that time, the young are getting a little large to herd, and they are beginning to forage off by themselves more and more, until the entire spawn is dispersed. Following a particular pair for over a month is difficult in the wild, and it is for that reason that whether the pair remains bonded and eventually spawns again is not known. Certainly, oscars sticking together when they are not in the spawning mode has not been reported.

While oscars have been aquarium favorites for many decades, they were discovered by science way back in 1831. The species was described by the famous ichthyologist Louis Agassiz as *Lobotes ocellatus*. There were several other cichlids included in the genus, but the genus was described so broadly that it was inevitable that future ichthyologists would modify it. The genus *Astronotus* (meaning "star-marked") was established by Swainson in 1839, utilizing Cuvier's *ocellatus* as the type species (i.e. an example of the genus). There were other species of *Astronotus* described, too, including *Astronotus orbiculatus*; however, many ichthyologists have tended to synonymize them. (That means that they have considered the name *Astronotus orbiculatus* as a junior synonym for *Astronotus ocellatus*). The species was described first as *ocellatus*, so that name has precedence according to the rules of nomenclature. After studying Haseman's original description of *orbiculatus*, Axelrod proposed utilizing that name for oscars that lacked ocelli (eye spots) in the dorsal fin and in the pectoral. Kullander, who has done much work to change the scientific names of South American cichlids, has indicated that there may be even more than two species of oscars, and the

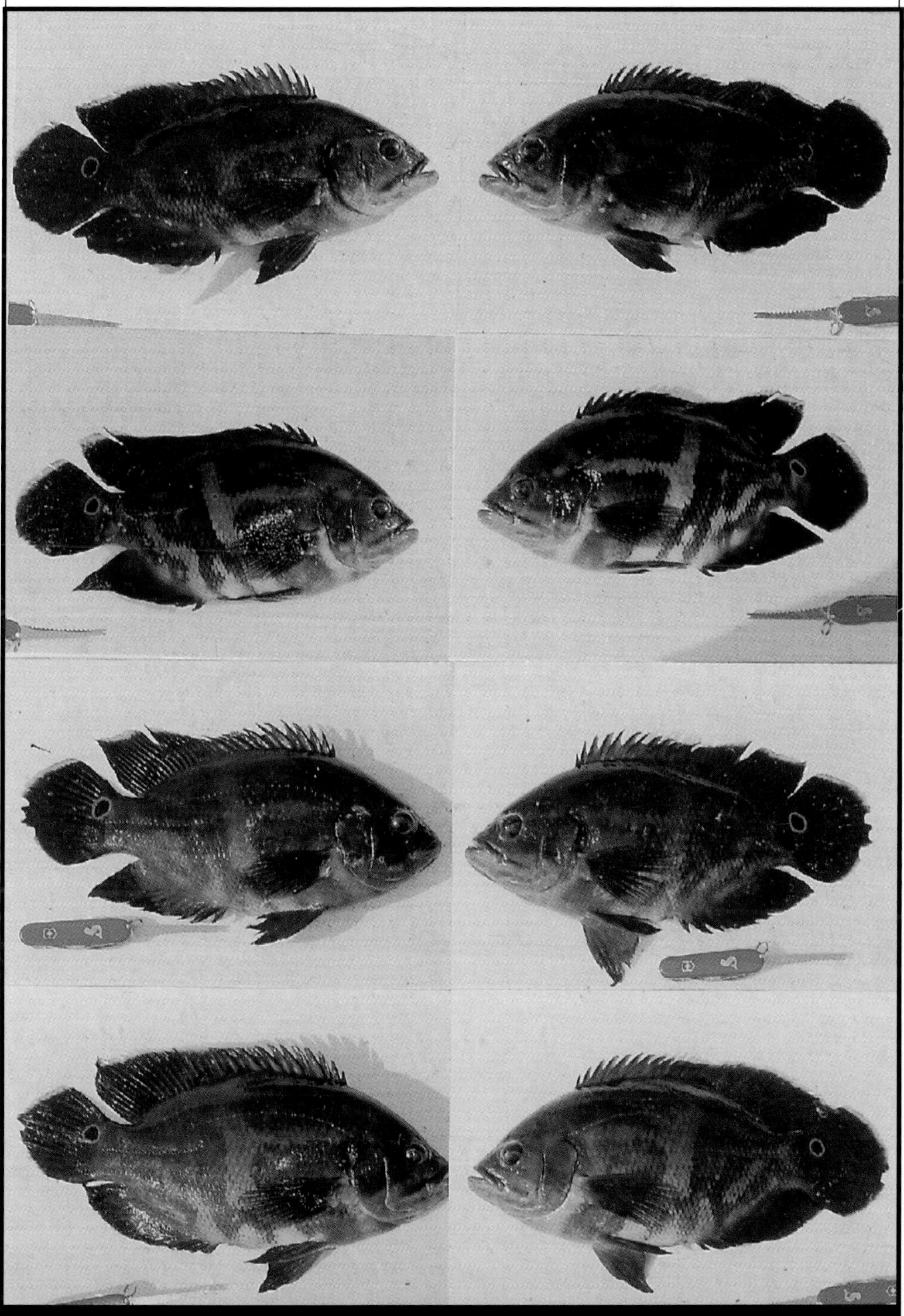

Oscars, *Astronotus ocellatus*, collected and photographed by Dr. Labbish Chao on the Rio Negro, Brazil. The opened knife is about 5 inches long.

Barely evident in freshly killed specimens from the Rio Negro are the ocelli (spots) in the dorsal fin. This is a typical adult specimen of the Rio Negro oscar, *Astronotus ocellatus.* Photo by Dr. Labbish Chao.

This specimen has been dead for 24 hours. It is an oscar from the Tefe region of Brazil. It clearly shows the ocelli (eye spots) along the dorsal and on the caudal peduncle (base of the tail). The spots vary from fish to fish and from side to side. The spots on the other side of this fish differ from the spots visible in this photo. Photographed and collected by Dr. Herbert R. Axelrod.

The ocelli on the caudál peduncle (tail base) are almost always fairly circular and distinct. The ocelli on the base of the dorsal fin are extremely varied in shape. This fish has many scales which have red centers. Photographed and collected by Dr. Herbert R. Axelrod.

The ocelli on this oscar are larger on the dorsal fin than on the caudal peduncle. Photographed and collected by Dr. Herbert R. Axelrod.

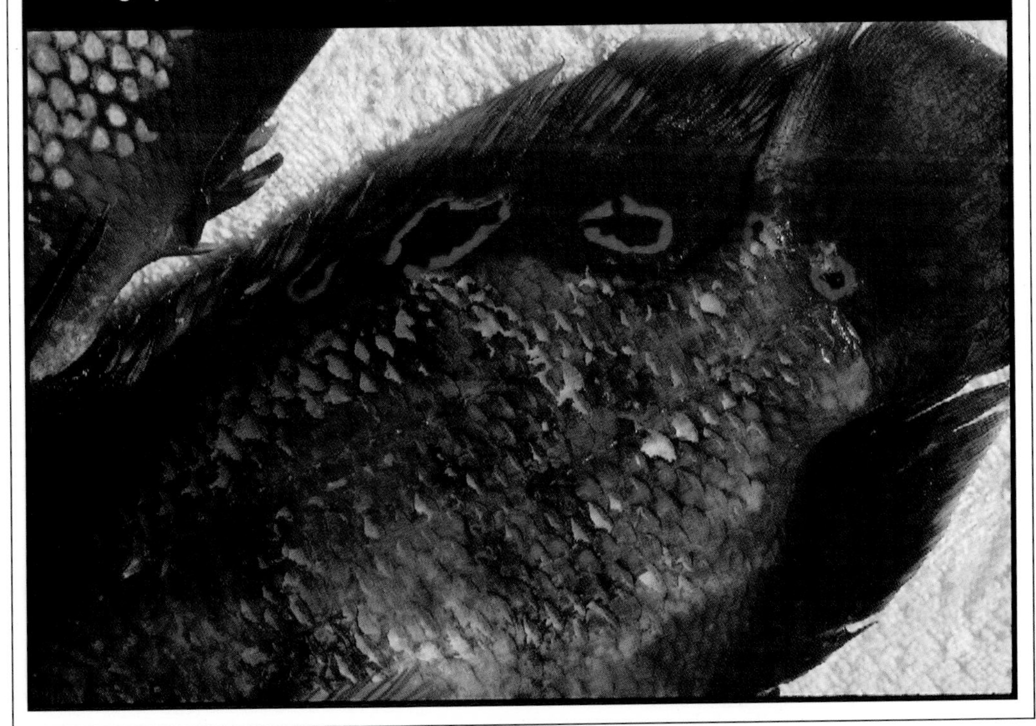

scientific and aquarium world await clarification.

In any case, the oscar is now known scientifically as *Astronotus ocellatus* (Agassiz 1831). The describer's name is placed in parentheses to indicate that his original description included the fish under a different genus (in this case, *Lobotes*). Whatever further taxonomic study brings, an oscar will still be an oscar, and very likely it will retain the same scientific name. A species that is as widespread as the oscar is certain to have, at the very least, geographical variations. Such variations are usually classified as subspecies. Since there have been collections many times from many parts of South America, we have had an opportunity to observe considerable variation in the coloration of the wild-type specimens. It may be, however, that we have some hybridization in the different color varieties that we have seen as aquarium species. Only time will tell.

Most hobbyists don't care that much about the scientific nomenclature or the life of oscars in the wild. However, I have known oscar lovers who made tourist trips to the Amazon and had to consciously restrain themselves from freeing the all-too-familiar oscars from fishermen's nets, as the fish obviously were in distress and were as beloved to these people as relatives!

The proof Dr. Axelrod offers for two species is the difference in the markings of the juvenile oscars from different river systems in Brazil. He collected these two juveniles when they were one inch long in the Rio Negro near Barcelos. He grew them up to maturity in his aquarium and noted continued differences. He calls this species *Astronotus ocellatus.* Photo by Dr. Herbert R. Axelrod.

The fact is that nearly all of the oscars that we find in the typical fish store are individuals that have been bred on fish farms or in hobbyists' tanks. Many of the new varieties have come from Singapore and places such as that. Such animals may no longer be considered a completely natural fish, so why not enjoy the different color forms? Even though such artificial varieties are anathema to a purist like me, I would be the last one to discourage the keeping of them or even the breeding of them. They undoubtedly are more forgiving of water quality and more easy to breed than the wild oscar. Even so, certain demented and devoted oscar lovers (such as myself) are always willing to pay extra for specimens that either came from the wild or are first generation (F1) descendants of wild fish.

There is something particularly noble and unspoiled about such fishes. They have a dignity and beauty that surpasses that of artificial varieties. And they have all their untamed instincts to allow us to learn from them, too. However, that is just the view of an admitted extremist. I am one of those handful of tropical fish lunatics who prefers my fish life on the wild side! And unlike many hobbyists, I would, given the opportunity, take my oscars straight from the Amazon—or the Orinoco, for that matter!

This is a juvenile *Astronotus orbiculatus* as identified and photographed by Dr. Herbert R. Axelrod. This is the non-acid water species of *Astronotus*.

LOBOTES. 129

laribus serie dentium acutorum exsertorum armatum, pone quos fascia lata dentium velutinorum minimorum exstat. Ossa pharyngea superiora et inferiora latis fasciis dentium magnorum obtuseconicorum apiceque subhamatorum armata. Linea lateralis interrupta. Pinna dorsalis longissima, antice aculeis acutissimis, apice cuticula appendiculatis, postice radiis articulatis sensim longioribus, et omnino squamatis suffulta. Pinna caudalis omnino squamata. Pinna analis antice aculeis validissimis aucta, caeterum eandem praebet indolem, quam pars mollis pinnae dorsalis. Pinnae ventrales acuminatae, aculeo primo breviore; pectorales subrotundatae.

1. LOBOTES OCELLATUS Cuv. in litt. Tab. LXVIII.

Aculeis pinnae dorsalis omnibus ejusdem longitudinis, excepto primo breviore; ocellis nigris alis marginatis ad basin pinnarum pectoralium pinnae dorsalis et in lobo superiore pinnae caudalis.

Caput crassissimum, obtusissimum, quartam partem totius longitudinis aequans. Operculum latum rotundatum, squamis illis trunci parum minoribus obtectum; praeoperculum laeve, alepidotum, magnis foveis mucosis notatum. Membrana branchiostega tenuis, lata, pone operculum prominens. Apertura branchialis sat magna. Bucae intumidae, squamis parvis obsitae. Oculi magni pone et supra maxillarum commissuram siti. Nares utrinque simplices, pyriformes, mediocres, inter oculos et rostri apicem intermediae. Plurima foramina mucosa, gregatim in summo vertice, ad nares, et in ossibus infraorbitalibus aperiuntur. Os labris crassis cinctum; mandibula crassa, latiuscula, prominens. Truncus latus, crassus, squamis subaequalibus, deciduis ad dorsi marginem et versus caudae apicem minoribus, tectus. Squamae omnes circulares, maxime regulares, lineis curvis concentricis confertissimis et radiis pluribus antrorsum divergentibus, et in marginem lobulatum decurrentibus notatae. Punctum radiationis in mediis squamis. Linea lateralis sursum flexa, cum dorso parallela, usque ad medium pinnae dorsalis posterioris partem, inde interupta, in medio caudae latere recta. Squamae ejus vix minores canali mucoso simplici, a media squama oriundo et in margine postico emarginato, aperto, insignes. Pinna dorsalis ultra medium aculeis validis, acutissimis, apice ut in Labris appendiculatis, postice radiis articulatis, gracilibus, apice pluries fissis caudae apicem versus sensim majoribus et per totam longitudinem squamulis minimis obducta. Pinna caudalis rotundata, omnino squamulata; extus radiis minimis cuti immersis vix conspicuis, caeteris radiis pluries fissis suffulta. Pinna analis antice aculeis tribus brevioribus, validissimis, crassis, acutissimis, radiis sequentibus ejusdem indolis et formae quam illi partis mollis pinnae dorsalis. Aculeus primus pinnarum ventralium, radio sequenti duplo brevior. Pinnae pectoralis radius primus brevis tenuis, parvus; sequentes radii plures et profunde fissi.

Dorsum virescens, venter et latera flavicantia.

Pinnae pectorales radiis 15, ventrales 6, analis 3, 16, caudalis intus 15, dorsalis 13, 20.

In Museo Monacensi specimen in spiritu vini servatur 10″ longum.

Habitat in Oceano Atlantico.

LEFT: The original description of *Lobotes ocellatus* by Cuvier became the synonym for *Astronotus ocellatus.*

BELOW: The drawing that was used to identify the original description of *Lobotes ocellatus* shows a fish with ocelli on the dorsal base and caudal peduncle. Axelrod followed this original illustration to differentiate between these two species.

CARE AND FEEDING

BASIC CONSIDERATIONS

One of the first things to consider in keeping oscars is what size tank to set up for them. The fact is that far too many oscars are bought and placed in ten-gallon tanks. This is okay if people know what they are getting themselves into and have very definite plans to purchase a larger tank at a later date. However, many people are clearly charmed by juvenile oscars, and they purchase one and place it in their community ten-gallon tank. They are in for a few surprises. Not only will their new acquisition grow at an amazing rate, it will begin to consume the inhabitants of the tank, which the hobbyist has been thoughtful enough to provide. Eventually, such people end up with a fish that is too large for their tank, and they may not always become such oscar converts that they are willing to purchase a larger tank for it. In such cases, the tropical fish hobby has lost in multiple ways. The fish suffers, the hobby has lost potential oscar devotees, and the people involved may just give up the entire tropical fish avocation in disgust.

A freshly killed specimen of *Astronotus ocellatus* identified, collected and photographed by Dr. Herbert R. Axelrod. This fish, from the Tefe region of Brazil, was the basis for the development of the red oscars.

Although I don't remember the exact date, I remember clearly the first time I saw juvenile oscars, some forty years ago. I had

become a rabid hobbyist and had a community tank at home. (Yes, it was a ten-gallon one!) My wife was enchanted by the tank of young oscars. I commented to one of the staff who was busy cleaning tanks not far from us that they were certainly fish with personality. He nodded in assent. "But," he said, "This is what they mature into." And he led us to the shop's resident pet oscar. Both of us gasped at the size of the animal. Not only was he big, but he was ramming the front of the glass to get the attention of the sales person. After all, an oscar is always hungry, and the ones that are kept by themselves are always seeking attention.

I wasn't interested in setting up a tank just for oscars at that particular time, but the point is that I had the information that I needed. As tolerant as my wife had been of my (then) new hobby, I would have purchased this very charming juvenile fish just for her. And, at that particular time when we only had a ten-gallon tank, that would have been a disaster! If we hadn't been warned about the size, a further temptation was that I liked them too.

We studied the oscar at the shop. This was the largest shop in town, and we had noticed the animal before. He and a large arowana (in another tank) were the display fish of the store. I kept going back to study the oscar. This was at a time before oscars had been bred for special coloration, so this animal was of the wild type, but he had a definite handsome appearance. This particular specimen had a lot of the green coloration in the anterior part of the body, with the typical black blotches on the side, along with the red pattern marks. Both my wife and I were struck by the personality and apparent intelligence of the fish and commented about it. The salesman nodded and remarked that they were probably the most intelligent of all fish.

In those days, we made frequent trips up to Los Angeles, as my in-laws lived there. I quickly became aware that there were also many fine fish shops there, and I haunted them with great frequency. One time I recall that a salesperson announced that he was going to feed a fish to a tank of oscars. He had the soon-to-be-eaten fish in his net, and he warned us that we would have to watch carefully, as the fish would disappear quickly. When the fish was dropped in, the oscars pounced on it with such rapidity that water was splashed out of the tank onto the floor.

There was quite a crowd of people there, and I am sure that oscar fans were both gained and lost that day. For one thing, I think the women present (and maybe even some of the men, too!) were concerned about all of this splashing of water. This need not be a concern if your tank is set up properly. Others, I am quite certain, found it distasteful, if not morally reprehensible, to be feeding live fish like that. Although I have pretty much come to terms with a fish-eat-fish

Dr. Schomburgk, in 1843, in his book Fishes of Guiana, II, plate X, illustrated another species of *Astronotus;* he called it *Astronotus rubro-ocellatus.*

world, I realize that many others don't like the idea of deliberately feeding live animals. It is one thing for it to happen in the wild, but in many people's minds, it is quite another for it to be deliberately perpetuated in "civilization." For those people let me assure you that an oscar can be kept in very good health with good growth without having had a single live fish in all of its days.

My wife was with me on one of my treks to the Los Angeles shops, and that was when we met a particularly affable oscar. The shop owner was perfectly willing to demonstrate just what a charmer this specific oscar was. He reached his hand in the water, and the oscar rolled on his side to invite being stroked on the side. So unusually friendly was this oscar that his owner could invite my wife to pet him, too. That did it! We were hooked, and my wife became my partner–in–crime instead of an obstacle to my plans of buying a large tank just for housing some oscars.

And that is the way it should be. We got our first oscars with some knowledge about them, and we purchased an adequate tank for them. I am not sure that we would have become the oscar fans that we remain today if we had gone the other route. I'm referring to the possibility that we might have contacted a different salesperson that first time and have ended up purchasing one of the juvenile oscars for our small community tank. Fortunately, sensible dealers fully inform their customers. To do otherwise is to do a disservice to them and to the fish involved.

THE OSCAR TANK

Most people start oscars out in a small tank. That is a perfectly reasonable thing to do as long as you have a plan in mind and an ultimate tank for the oscar to be kept in when it has attained some of its prodigious size. The question is how big should the tank be and what kind of filtration should it have. Let me give some recommendations based on nearly forty years' experience with these remarkable animals.

First, the size of the tank will depend to some degree on your expertise and diligence as an aquarist. If you are willing to make weekly or even daily water changes, you can make do with a smaller tank. Also, you need to take into account whether or not you are going to keep one oscar or several. Presumably, you will only end up with a pair of oscars, but if you are going to keep a community tank of oscars, you will need at least a 200-gallon tank. We'll talk more about the logistics of all of this later on.

The fact is that you can make do with a 30-gallon tank; however, a 50 is better and a 125-gallon tank is preferable, and this is especially so if you intend to keep a pair of these fish. A larger tank allows you a margin of error, and it certainly will make your oscar (or oscars) happier. Oscars are excellent candidates for keeping in a big tank, as they tend to hold a territory anyway, but a little extra room is appreciated.

DECORATIONS

Most cichlids, including oscars, won't eat plants, they will destroy them. Part of this is just aggressive behavior, and part of it is part of the spawning routine. Even if an oscar is kept by itself, it will still display some spawning behaviors. In fact, large females have been known to clean out the bottom of a tank and lay eggs all by themselves. In any case, oscars, along with many other cichlids, like to uproot plants at spawning time. This behavior probably has a purpose, as it gives the spawning fish a clear view around them and doesn't allow any hiding places for predators to lurk. But spawning cichlids have a lot of built-in aggression which they use to good advantage in the wild to ward off would-be predators of their eggs and young. However, in the tank, it can cause problems in that the fish will "attack" plants and even filters and heaters. This is technically called "displacement behavior," and that term refers to the fish deflecting its aggression to plants or inanimate objects. Whatever the cause, plants are not recommended for an oscar tank. At least, that is the way I have always done it. I must confess that I have known some hobbyists who were successful in keeping some very tough *Vallisneria* plants in the deep sand in the back of a large tank,

According to Axelrod the oscar without any ocelli on the dorsal should be identified as *Astronotus orbiculatus*. Photo by Wolfram Ch. Schrey.

but I suspect a lot of re-planting was involved.

So the question is what to use for decorations in the oscar tank. Well, naturally rocks can still be used, but some thought should go into selection of these, also. Oscars are famous for moving sand around and even picking up smaller rocks. You need to have large rocks and make sure that they are on the bottom of the tank, not just on top of the gravel. Manufacturers have come to the aid of oscar and other cichlid enthusiasts in this respect, as many of the ornaments sold consist of artificial rocks and drift-wood. These have the advantage of not being as heavy as the real thing, and there is no danger of contamination from them.

Many hobbyists, however, prefer real rocks and driftwood. Rocks of volcanic origin are safe, but driftwood should be soaked for many weeks in a container in which the water is changed regularly. After the driftwood is no longer turning the water an amber color, it should be safe to use. Still, to be completely safe, it is best to utilize an inexpensive "test" fish before placing this wood in your tank.

Truth to tell, oscars don't need a lot of rock and driftwood, as so many other cichlids (and some catfishes) do in order to feel secure in their tank. One large rock is fine, but most people like to have the tank decorated in some way. One thing that you can do here is to utilize artificial plants. They may not grow, but they look like the real thing, and they don't die. If you are setting up a tank for oscars or other cichlids, you can utilize a little silicone cement to anchor the plants to the bottom of the tank. Do this before you put the gravel or water in the tank. Set up the plants in the way and the location that you like them, then anchor them with the cement. Don't spare the cement, as the oscars are sure to do at least a little tugging on these plants.

As for gravel, I have kept oscars with everything from fine gravel to very coarse gravel (including shattered brick fragments), all with equal success. Nevertheless, I would suggest medium-sized gravel. It has two advantages. It doesn't harbor as much debris as coarse gravel and is easier to "vacuum" than fine silt, which will tend to enter the vacuum or special adaptor on a siphon hose. One thing as certain as death and taxes is that your oscars are going to be moving the gravel around, so you won't have to worry about dead spots developing in the gravel in which anaerobic bacteria may grow.

FILTRATION

One of the first things to eliminate here is the undergravel filter. Since oscars are large and have prodigious appetites, it only stands to reason that they are going to be producing more waste products than a tank of tetras. For that reason, an undergravel filter is going to be quickly overwhelmed, and, unless you clean your gravel almost daily, it is going to do more harm than good. Another problem with the undergravel filter is that oscars dig constantly. They will remove the gravel from the filter plate, and this considerably reduces the efficiency of the filter.

There are several possibilities for filtration, and I have seen many setups that worked, including a tank with a bunch of box filters. This is the economical approach, but the filters must be changed weekly at the very least, and the oscars may drag them around the tank and pull on the air lines. Sometimes they pull the air stones out of the filters and chew them up. Yes, oscars may be as much fun as dogs, but sometimes they can be almost as frustrating, too! The point is, though, that such a simple system can work. The key to making it work is to first change the filters regularly and change the water at least weekly. With such a system, many hobbyists make about a ten percent daily water change. Others change about twenty-five percent weekly. Even with other filtration systems, you will want to make regular partial water changes, even if they aren't of such a great magnitude.

Nutritious foods for aquarium fishes of all types are available at pet shops and tropical fish specialty stores. Flake foods such as those shown here are among the most popular of the different forms produced commercially. Photo courtesy of Ocean Star International.

A combination of filters that I have found excellent is to use a combination of mechanical and biological filtration. This can be done with a canister filter to perform the mechanical filtration. They can be either of the cartridge type or one that holds filter floss and activated carbon. I prefer the cartridge type myself, as I plan on changing the mechanical filter regularly. For the biological filter, it is difficult to beat a fluidized bed system. These are of a convenient size and can hang on the back of the tank. They do a great job of breaking down ammonia into nitrites and then nitrates, which fish can more easily tolerate. These filters give you some leeway on your partial water changes, but it must be emphasized that one of the great secrets to success with oscars consists of regular partial water changes. But it helps to have good equipment, too.

Whatever filtration you utilize for mechanical filtration, I would recommend a diatomaceous earth filter as a supplement. This can be utilized while you are doing your partial water changes. That is, it can be used on a weekly basis to "polish" your water by cleaning it of all debris. These filters are not recommended as a primary filter for an oscar tank as they are too efficient. This may seem a paradox, but the problem is that the filter becomes clogged with just the protective coating from the bodies of the very large fish that you are keeping, and the sleeves that hold the diatoms must be flushed out and cleaned almost hourly.

Another alternative is to utilize two canister filters. That way you can set up the canister filters so that there is biological filtration in them. If you change one filter a week, you are allowing the other one to build up the de-nitrifying bacteria in the alternate filter. In the meantime, the newly-cleaned canister filter functions as a mechanical filter, while its "twin" takes care of the biological filtration (along with its own mechanical filtration).

Finally, it is possible to utilize trickle filtration to help maintain water quality. Such filtration is primarily intended as biological filtration, but most such units have a prefilter, which is a mechanical filter that should be cleaned daily. If you want the ultimate in filtration, you can combine fluidized bed filters with a trickle filter system. Although both of these filters are biological, they complement each other very well. This is because the emphasis is on gas exchange in the water with trickle filtration, while a fluidized bed unit maximizes de-nitrifying bacterial activity.

So we have gone from the economical setup of inside box filters to a system of compound filters that includes canister filters, a fluidized bed filter, and a trickle filtration set up. Obviously, the latter configuration is going to be a lot more expensive. Is it essential? No, it is just nice to have, and it saves you some degree of work, as it maintains the water quality for a longer period of time. (With the box filter system, we were not relying on biological filtration at all; we circumvented water deterioration problems by daily partial water changes.)

Even though you are not using an undergravel filter, you will need to clean the gravel. Perfectionists occasionally siphon out all of the gravel and flush it in a pail until the water runs off clear. However, most of us can get by quite well by cleaning the gravel with a special adaptor on the siphon hose which allows us to clean the gravel without siphoning it out. Those who love gadgets and special convenience may wish to make use of the powered water vacuum cleaners which are sold by dealers. These are certainly handy devices, but we must remember to make the regular partial water changes and not be fooled into complacency just because the water is clear.

LIGHTING

Since you are not going to be keeping plants, personal preferences can pretty much hold sway here. I always preferred full-spectrum fluorescent bulbs to the ones which emphasize the reds (which is the case with many lights sold with aquariums); however, that is a personal preference on my part. Most oscar people I know seem to feel that their pets preferred subdued lighting, but I have seen oscars thrive in brightly-lit tanks too, so it is again pretty much personal choice on this. Some people like to keep a surface of floating duckweed. I have never seen oscars eat this except by accident. Nevertheless, many hobbyists like it because they feel that it helps in water quality problems. Such an idea is not without merit, as plants provide biological filtration, too, in that they utilize some of the nitrates produced by the fish.

The point is that if you do decide to utilize duckweed, you will need sufficient lighting for it. However, the normal fluorescent bulbs that are standard equipment on most tanks seem to provide sufficient light. One of the drawbacks of duckweed is that any time that you place an arm in the tank (to clean the front glass, for example), you will have to rinse the tiny duckweed off that inevitably clings to your arm after you remove it from the tank.

The large mouth of this oscar indicates its ability to swallow small fishes in one gulp. It does not have cutting teeth like a piranha so it cannot take a meaningful bite from a larger fish. Photo by Burkhard Kahl.

WATER CONDITIONS

Although wild oscars tend to prefer water on the soft and slightly acidic side, captive-bred oscars tend to be extremely tolerant of pH differences and hardness. Truth to tell, many of the oscars from the wild are going to be from widely variant water conditions, too, so the point is that it is not of paramount interest that you establish a specific pH and hardness in the tank. The more important thing here is to keep the water free of metabolites. So just keep the water as it comes from your tap (in regard to hardness and pH, of course, as you will want to use water conditioners to rid the water of chlorine or chloramine).

COVER FOR THE OSCAR TANK

Fortunately, most aquaria come with adequate covers. Remember the oscars we mentioned at the fish shop that splashed water out on the floor when they were fed live fish? That is more likely to happen when you have more than one oscar in the tank and they are thereby competing with each other for any food that is placed in the tank. That makes them even more rambunctious at feeding time. Still, a cover helps out a lot here, regardless of the situation you have. Some people prefer their tanks more tightly covered than others do. With oscars we don't have to worry about small openings at the front or back, as these big fish aren't going to be able to jump through them as small fish might. Still, I prefer any gap that exists to be in the center of the aquarium, right down the middle. I do everything but my cleanups through there. That obviates any problem of water splashing out when I feed the fish.

It is safe to have such a gap down the middle, as fish that jump out of a tank nearly always go up the side and over. In any case, such a small gap is not going to leave room for oscars to get through anyway. And in addition to all of that, it is nearly impossible for a fish to jump out of the tank by going out through a gap down the center.

Some hobbyists are much more fussy than I am, and they prefer to have the tank tightly covered in order to cut down on evaporation. That way they don't have to replace evaporated water as often, and the tank will not contribute to any humidity in the house. Since I live in an area with an ideal climate, such concerns are not important to me, and I like easy access to feeding the fish. Besides, I like the circulation of air that such a gap allows, as it facilitates gas exchange at the surface.

SPECIAL CONSIDERATIONS

Oscars can sometimes be a real pain about attacking such things as filter outlets and intakes, along with the heater. As mentioned earlier, oscars are of individual personalities, and some are worse than others. In fact, with some oscars, this is hardly a problem. Attacking the heater is a particular problem, as it is possible that the oscar may crack it, creating a real catastrophe, or knock it clear out of the tank. If you are using a trickle filter, you can simply place the heater in the filter. If you don't have that option, you need to think in terms of a partition to keep your oscar away from the heater or anything else it may be damaging. A good partition can be provided by utilizing light-diffusing grids from large fluorescent lighting fixtures. Just cut it to the size that you need and use either rocks or silicone cement to keep it in place. I have never had a problem with toxicity with these things, but I recommend testing them with an expendable fish before you place them in your oscar tank. (Oscars most definitely are not expendable—especially when you have had one particular individual for over ten years!)

In the developmental history of the red oscar, fishes appeared with more and more red in their scales. The variety was not genetically fixed (reproducible) until the fish became almost totally red.

MAINTAINING THE TANK

Whatever filtration system you have selected, you will still need to do water changes. The question is how many and how much. Again, this is an individual choice. Some hobbyists enjoy fussing with the tank and doing water changes. To others, it is more of a chore, one that they prefer to put off as long as possible. If you are in the latter category, you will want the multiple filtration (or compound filtration, as some call it) setup, as that will allow you to delay longer between water changes. In such cases, you will want to have test kits for ammonia, nitrite, and nitrate, as they will tell you when you need to make a partial water changes. Any buildup of ammonia or nitrites is an indication that a partial water change is necessary.

One of the things that will make water changes less onerous is a good water delivery and automatic siphon system. These are available from fish shops, and they should probably be purchased along with the tank and other tank equipment.

Cleaning the glass is largely a matter of taste. I always use background paper or dioramas in my display tanks, as such things help to show the fish off to better advantage. I don't worry about algae growing on the back and sides (unless I have a really nice diorama on the back that I don't want a growth of algae to obscure), but I keep the front glass crystal clear. Since there are occasions when I may not want to place my arm in the tank, I use a pad with an extended handle on it. However, I also have a pad that is hand-held for those occasions that I am feeling particularly zestful about going into the tank and cleaning all nooks and crannies of algae.

The cover glass will occasionally need to be removed and cleaned thoroughly; however, this is a rare chore. The most fussy and immaculate of aquarists is not going to need to do that more than once a month. The ornaments and driftwood may occasionally need to be cleared of algae, but this is something I leave alone. (However, I nearly always have a plecostomus-type catfish in my oscar tanks, too.)

The fact is that oscar tanks are relatively maintenance-free except for the regular partial water changes.

TANKMATES

Obviously, there is a balance to be found here. You don't want tank-mates that the oscar (or oscars) will regard as food, and you don't want fish which are likely to pester the oscars. The fact is that some smaller cichlids can give oscars a bad time, as oscars are not really extremely aggressive as compared to many other cichlids, such as Texas Cichlids, Red Devils, Red Terrors, Green Terrors, and a host of others. In fact, I am hesitant to suggest other cichlids as tank-mates, although a few severum would probably be okay because they are too big to be eaten, and they aren't any more aggressive than the oscars.

The armored catfish known in the hobby as "plecos" (short for *Plecostomus*, which used to be a generic name for some of these species) are a good tank-mate for oscars. That is because these fish will rarely bother each other. Plecos are normally nocturnal, although they will come out in the daylight to feed. In any case, they pretty much stay hidden and inactive during the day. Even when they are out, their armored bodies will protect them from the oscars; however, you will want only one to a tank, as they will fight with each other, and all of this is likely to take place at night when you won't see it. Beauty is in the eye of the beholder, but many people find these very odd–looking fish to have a certain elegance. They are certainly unusual in appearance, and they can be

The final stage in the development of the red oscar. Fishes are now produced which are extremely similar to this one. Photo by Andre Roth.

helpful in that they will utilize their night-time hours to scour the tank of algae.

Some very good tank-mates for oscars are silver dollars. They get large enough and are maneuverable enough that the oscars are unlikely to hurt them, and they provide a nice contrast, both in appearance and in manner, to the oscars. Oscars are of darker hues (unless you have albinos) and silver dollars are, of course, silvery, and they have a laterally compressed shape, as opposed to the fullness of the oscar. The continual swimming around the tank by the silver dollars not only assures constant movement in the tank, it provides a certain assurance to the oscars that everything is normal and fish are out and about. Also good tank-mates are tinfoil barbs and possibly even pacus, but pacus get even bigger than the oscars. They are not a threat to the oscars, but you should be aware of how big they get. And bigger than an oscar is big!

If you plan on breeding your oscars, it is best if any tank-mates are removed well before breeding time. You don't want any fish in there at breeding time, as the pair of oscars will direct considerable aggression toward them (and if they survive the attacks, they may eat some of the eggs and babies). If you attempt to move such fish when the oscars get in a spawning mood, the hullabaloo involved may throw them off the spawning track.

The bulging eyes of this *Astronotus* do not indicate exophthalmia (pop-eyes) but are normal for this species.

A HEALTH PLAN FOR YOUR OSCAR

Fortunately, oscars are basically healthy fish. And the best health plan for your oscar is prevention. Provide your fish with good quality water. Good filtration and conscientious water changes will take care of that.

Also, be sure to quarantine any fish that you decide to keep with your oscar; that includes potential mates. Sure, it is a bother, but it is a lot more bothersome to be tending an entire tank of sick fish that didn't have to be that way. Set up a separate tank for quarantining. It doesn't have to have exotic filtration. You can compensate for that with copious water changes. Observe your charges there for at least two weeks before an attempted introduction into your oscar tank.

If your oscar does come down with ich or velvet, the standard medications work quite well with them. They are good patients, and they mend quickly.

The only disease to which oscars may seem particularly vulnerable actually has to do more with the oscar's size than the fact that it is vulnerable to the disease. The disease in mind is "hole in the head" disease, and it refers to the erosion of areas of the skin around sensory pits, most particularly along the lateral line. This same ailment is the bane of marine hobbyists, too, although it is much less likely to happen with oscars.

There are differing opinions about the cause of the ailment, although many experts suspect the pathogen *Hexamita*, a protozoan that is often present at the site of the pittings. It is now widely believed that stress from poor water quality is the major cause of the ailment. So make those water changes and be patient, as many an oscar has recovered from this unsightly ailment. That it rarely seems to do harm is the consensus among marine hobbyists.

The main lesson here is "tender, loving care," for many hobbyists keep fish all of their lives with almost no fish contracting any ailments. They accomplish this seeming miracle by good aquarium practices and avoiding stress for their charges.

FEEDING YOUR OSCAR

First, you have to decide if you are going to spoil your oscars and feed them a lot of live food, specifically fish. (Crayfish can be utilized, too, if you have access to a lot of them.) The thing that has built the feeder goldfish industry more than anything else is the fact that people like to get them for their oscars—although they are also purchased for other cichlids, predatory catfishes, and marine fishes, such as lionfish. The truth is that feeder goldfish may be quite inexpensive, but they are not always a bargain, as they may very likely be carriers of disease. Quite frankly, I am not aware of any oscars coming down with any ailments from eating goldfish, but I am being hyper–cautious here.

If you do want to make a habit of feeding your oscar live food and especially fish (which, after all, is their natural diet), you may be able to make connections with some guppy or goldfish breeder and arrange to take his culls. An interesting fact here is that some goldfish or guppy breeders keep their own oscar as a pet and for just such a purpose: to dispose of the culls.

For a variety of reasons, a lot of people may not want to feed their oscars live fish at all. Or they may wish to save them for only an occasional treat. If that is the case, you should begin feeding your oscar dry food and other staples from the time that they are young. It is certainly true that oscars which have been fed only live fishes may refuse anything else until you get them really hungry (which is very difficult on oscar lovers!), so you want to start them out on other foods early. Oscars are especially eager eaters when they are young, so it is easy to train them to eat just about anything which hits the water at this stage.

Of course, there are other live foods besides fish. Young oscars can be given live brine shrimp occasionally. However, they will eventually get so large that brine shrimp are hardly worth the bother to them. At this stage, they can easily take whole earthworms. You don't even have to dig these yourself, as you can buy them at bait shops.

Mealworms are also a possibility, and these you can raise yourself. Simply fill a five-gallon tank with corn meal and place a few little slices of potato or apple in the top and place some mealworms in your culture. You will probably want a screen that fits over the top of the tank, as mealworms will climb out, especially after it gets really crowded in there. Besides that, mealworms are actually not worms. They are the larvae of a beetle, and you probably don't want them around your house, looking for some flour or something to lay their eggs in. Some hobbyists simply place a bit of denim over the top of the container (which can be a bucket as well as an aquarium), and that keeps the culture contained. It only takes about six weeks for your mealworms to have gone through the cycle, become beetles and reproduce, in order to provide you with a nice culture of mealworms.

Bits of shrimp and beef heart can also be fed to the oscars. In fact, beef heart was long the staple diet that most hobbyists used for feeding oscars. With the emphasis on a less meaty diet, it seems that fewer people are feeding their fish beef heart. Certainly, it is not a complete diet for the fish, but the fatty aspects of it are very likely not harmful for oscars. As predators, oscars are used to lots of protein and some fat, and I have seen some really large oscars that attained that size on a diet of largely beef heart. One of the advantages to beef heart is that it is lean and cheap.

Whatever you feed your oscars, there should be an emphasis on variety. We know that those hobbyists who fed their oscars only beef heart were probably mistaken to have done so. But even though oscars primarily eat live fish in the wild, they also nibble at other things and even ingest a little algae. A marked distinction from the early stages of the fish hobby is that dry fish foods are extremely nutritious these days. They are so good that even those who have their oscars so spoiled that they won't eat anything but live fish tend to make sure that the feeder fish are stuffed with dry food to make sure that their oscars get their full complement of vitamins!

So even if you choose to feed your oscars an assortment of foods, like mealworms, night crawlers, brine shrimp, and an occasional live fish, you should include some dry food in the diet. I have found that flakes work well with juvenile oscars, and I gradually work them up to pellets and even increase the size of these as the oscars get larger. The fact is that oscars are perfectly capable of thriving on a diet of dry

The development of the red oscar proceeded to produce fish which were more and more red. Photo by Burkhard Kahl.

food alone, but most of us vary such a diet with mealworms, earthworms (night crawlers), and maybe an occasional fish.

HOW OFTEN SHOULD THE OSCARS BE FED?

When to feed is almost as important as what to feed. The fact is that even juvenile oscars can get by on a feeding in the morning and one in the early evening. The really hard fact is that juvenile oscars can get by on just one feeding a day, but they won't grow nearly as fast, and the hobbyist will feel guilty. Oscars have a way of getting their way with us, and what they usually want is food. For that reason, oscar lovers are in more danger of overfeeding their fish rather than the other way around. The question is: can it be harmful to overfeed an oscar? In my experience, these incredible animals can eat just about everything but the kitchen sink (and probably even that, too, if they get the chance) without harm. But of course we always want to make certain that we are not polluting the water in the tank by feeding too much. So don't keep feeding your oscar once he is no longer ravenously bolting the food down.

The next natural question is: should oscars that we are attempting to get to breed receive a special diet? And that is a question that we will take up in the very next chapter.

BREEDING OSCARS

PAIRING OFF

If you become sufficiently enamored of oscars, you may very well decide that you would like to breed them. Of course, this is a misnomer. Cichlids aren't manipulated in a spawning situation as other fish are. That is, they don't scatter the eggs or lay them in spawning mops and leave them for the aquarist to harvest. It is for that reason that some of the greatest breeders of all time, such as Gene Wolfsheimer, have emphasized, "You don't spawn cichlids. Cichlids spawn themselves."

By that, Gene meant that it was the job of the aquarist to produce the right conditions for spawning. He knew well that doing this very thing could be difficult, so he was well aware that some cichlids were very difficult to get to spawn in the aquarium. And Wolfsheimer knew whereof he spoke, as he was one of the first people to spawn discus back in the days when it seemed nearly impossible to do.

While oscars are not difficult spawners, they are not easy either. Some hobbyists might think that they are, for many aquarists move from one species to another, right after they have spawned a particular fish species. Thus, they may have lucked out and obtained a pair of oscars which were easy spawners and thereby concluded that oscars were easy to spawn. The fact is that they can be difficult to pair, and even if you have a pair that is compatible, they can drive their keeper crazy with false starts or by just going long periods of time without exhibiting any reproductive behavior whatsoever. Every oscar hobbyist will know exactly what I am talking about in this particular instance.

Oscars are not difficult to induce to spawn if they are properly fed, have lots of room, and a suitable base upon which to deposit their eggs. This male is scrubbing the area upon which the eggs will eventually be deposited. Photo by Joe Kislin.

The first problem, of course, is in getting a pair. The easiest and most economical way is to obtain from six to eight juvenile oscars and allow them to pair off on their own once they have attained sexual maturity. This method calls for more patience than many neophyte hobbyists possess, for it can take a year to nearly two years before oscars are fully mature sexually. The alternative to this system can be expensive as well as frustrating.

A half-grown *Astronotus ocellatus* clearly shows the single ocellus on the caudal peduncle. The body striping varies from individual to individual and changes dramatically with moods. Photo by Burkhard Kahl.

To begin with, male and female oscars look the same. It is possible to ascertain the sex of oscars, but it involves netting these large animals out of the tank and holding one of these struggling behemoths in a net while utilizing a flashlight to closely examine the genital area of the fish. After having seen this done and having done it myself a number of times, I am undecided as to whether it is more traumatic to the fish or its keeper! It takes a little practice to tell the difference, but the male's spawning tube points back toward the tail and it is located just in front of the opening of the anus. The female spawning tube is less pointed and points straight down. For this reason, the female's genital area gives the impression of two openings (including the anus), and this method of determining the sex of certain fish species has sometimes been called the "one–hole–two–hole" method.

Even knowing the sex of the animals doesn't ensure that the fish will accept each other. The fish may reject one another or, at the very least, the courting period can be rough. Naturally the tank should be large (of at least a hundred gallons capacity), and the easiest approach may be to place a glass partition between the fish. You can have plate glass cut to fit your tank, and some dealers even supply glides that hold the glass in place. More likely, you will have to utilize rock work to hold the glass in place.

Whatever the situation, you should use a couple of pieces of slate, or some similar flat object that is safe to place in the water, to prop the glass up from the bottom about half an inch. They should be placed on the sides at the bottom, as we want to allow water circulation through the center area. The gravel should be cleared away from the area around the glass, as the fish are going to be digging to get at each other anyway. If you already have the gravel removed, you can accurately judge the distance beneath the glass and thus be certain that the opening is not large enough for one of the fish to squeeze underneath.

The fish can be kept together indefinitely with the glass between them. Even if the pair spawns, the female will lay the eggs as close to the male as possible, and his efforts to fertilize them will be surprisingly successful. The circulation of the water beneath the glass will be sufficient to allow the spermatozoa to reach most of the eggs. The female will then tend them, and she will pick out any eggs which have gone bad. And it should be mentioned here that oscar eggs look very much like normal cichlid eggs which have grown fungus, as they are unusually white in appearance.

Young fish about two inches long indicate they are likely to be *Astronotus ocellatus*. Photo by Harald Schultz.

Because of all that is involved in sexing a pair of adult oscars and then having to keep them with the glass partition between them, mated pairs of oscars are especially valuable. Expect to pay more for a mated pair than you would for just two adult oscars. There are a couple of pitfalls, however, even with mated pairs.

One is that the fish which had been so compatible all of these years may resort to fighting when they are placed in a strange tank. This is unusual, but it should be known that it can happen. The other pitfall is that sometimes two females will pair up under the artificial constraints of an aquarium, and they will even lay eggs together. Again, this is unusual, but it has happened. And when it has happened, invariably the individual who has them is certain that he has a mated pair of oscars. The eggs simply went bad or the fish ate them, or whatever. It is wise to get an agreement that you can return the fish in the case of either one of these eventualities.

While it may require more patience to raise your own oscars for breeding, it works better, and you can keep the fish together without the partition. To be honest, there are pitfalls here, too, as there are in nearly any situation imaginable. First, there is the extremely remote chance that you will get individuals of all one sex. In my experience, the chance of this happening is small to the vanishing point, but it is a possibility that must be considered. Then, too, you are going to end up with some extra oscars which you will need to find homes for. This is not usually a problem, as dealers are willing to take in adult oscars, as there always seem to be buyers for them. Even if oscars have been raised together and allowed to pair up naturally, there can be a sudden outbreak of hostilities, and you can end up with a pair with "irreconcilable differences," but, again, this is rare. An oscar pair gets along sufficiently well that it is evidence for the possibility that they pair for life in the wild.

Among the many advantages of raising your own oscars is that you get to know the animals better. You have observed them throughout their growth and their pairing activities. You can pretty much predict their behavior. It should be mentioned that several oscars, or even a pair, kept together are not quite as personable to their owners as is the case with just a single individual kept by itself. Even so, the personality of the fish shines through; it is just that you are dealing with a "couple" instead of an individual.

FEEDING POTENTIAL SPAWNERS

Unlike barbs, tetras, catfish and so many other aquarium fish, we are not trying to feed oscars with an unusually rich food supply to condition them for spawning and then place them in a separate tank for them to do their thing. As mentioned earlier, cichlids spawn themselves. However, the hobbyist can make the cichlids more likely to spawn, and that is true of oscars, too. The number one priority is good, clean water, but you should have been providing that all along.

Similarly, you should have been providing a variety of foods for your oscars, too. Then they will spawn almost automatically upon the onset of sexual maturity. If you haven't been feeding the pair well beforehand or someone else had one or both, now is the time to be really conscientious about providing your oscars with good quality food. The base of the diet should be a nutritious dry food, and there are some types which are formulated especially for cichlids, and these are excellent for oscars. Frozen foods, such as krill, and special formulations for cichlids are excellent, too. You can build on this basic diet by feeding occasional earthworms and mealworms. In an attempt to induce spawning, you may want to offer these daily but on alternate days. Also, an occasional feeding of live fish can be afforded the fish at this time. If you don't want them to get spoiled and refuse other food, make this offering not more than once a week. The truth is that they don't need live fish in their diet to induce them to spawn. But it most assuredly won't slow things down either.

In my experience, twice a day feedings are more than adequate for a pair getting ready to spawn. Some people have fed more often with good results, but most oscar keepers that I have known have gotten very good results by feeding only once a day. After all, in nature they may go several days without eating. (And on fortuitous days, they may feed several times in one day.)

THE ACTUAL SPAWNING

As was mentioned earlier, oscars can be maddening to their owners when spawning time

***Astronotus ocellatus*, a well pigmented tank-bred variety which clearly shows the ocelli in the dorsal fin. Photo by MP & C Piednoir Aqua Press.**

The red oscar is almost completely covered with red scales except for the head. Photo by Andre Roth.

arrives. There may be several false starts and many weeks of courtship and the mimicking of the spawning act itself before it actually happens. After all this wait, the oscars may confound their owners when they finally do spawn, as they may eat the eggs or, worse yet, the wigglers. Since most large cichlids are long lived, attaining at the very least ten years in longevity, it is believed that there has been more room for error with them. For that reason, there may be a learning period with these large cichlids. There may be a couple of spawnings which end in disaster. That doesn't necessarily mean that the pair is behaviorally defective. After some aborted attempts (two or three), the pair may spawn once again and thereafter be model parents.

Although the possibility of a defective pair must eventually be faced, a little extra leeway is recommended in the case of oscars. Naturally, the eggs can be reared artificially. This can be accomplished by placing the eggs in a small container and placing an air stone near them to establish a flow of water over the eggs to assure the natural water flow over them that is usually provided by the parents. Another possibility is to direct a flow of water over the eggs with a powerhead. Many hobbyists also like to dye the water with methylene blue, but the most important thing for the eggs and eventual wigglers is water flow over the eggs and good clean water that is well oxygenated. The fry will become free swimming in about four days after hatching, and they will greedily take newly-hatched brine shrimp at that time. Live brine shrimp is the best, but the frozen version will suffice. Small bits of dry food of the flake type can be placed in the tank, and they too will be taken even near the first day of the free swimming stage.

Now having said all of that, let me interject a personal note here. I am opposed to the artificial hatching of oscar eggs. Part of the charm of oscars is the devoted care that they give to their young. It would be a shame if that were lost among much of the captive breeding stock because the progeny of behaviorally defective parents were being raised by breeders. In the wild, the genetic material of such animals would be unperpetuated. I think that we should be true to nature on that score.

Besides, it is fun watching oscar parents with their young. Each parent devotes equal care. When a youngster strays too far away from the shoal, one of the parents scoops it up in the mouth and spits it back into the center of the group. The fry flock around the parents and they are attentive to certain signals given by the parents. For example, a flicking of the ventral fins by the parents will cause the fry to hug the bottom, as apparently that is a signal that is given when danger is about. Also, when you feed the parent fish pellet food, you can enjoy watching the parents pulverize the pellet and spit some of it out for the young. In fact, you may have to cut back on your feedings to your oscars, as they become more messy feeders when they are tending young.

If you do things the natural way and allow your oscar parents to tend their young, you most certainly will have a more instructive and enriching experience. Not only that, but the young fish will prosper better, too. Many cichlid fish that are kept with the parents grow much better than the ones which have been separated. This is believed to be due to the fact that many cichlid parents secrete a specialized body slime upon which the fry can feed. In most cichlids it is a supplementary food, but it is obviously important in that they grow so much better in the presence of the parents.

The fry should be removed at about four weeks of age. It is believed that this is the age that they begin to disperse in the wild. An easy way to remove them is simply to siphon them out with a thick siphon hose into a plastic bucket. The larger the "grow-out" tank for the fry is, the better off they will be. Make sure that you have the water in the tank that is receiving them at the same parameters as the water they are coming from. (Okay, cleaner is all right, but be sure that the pH and temperature are the same.)

The fry should be ready to sell to your dealer in another four weeks. And you can pass out cigars among your fellow tropical fish hobbyists. You have joined the ranks of those who have helped oscars reproduce themselves.

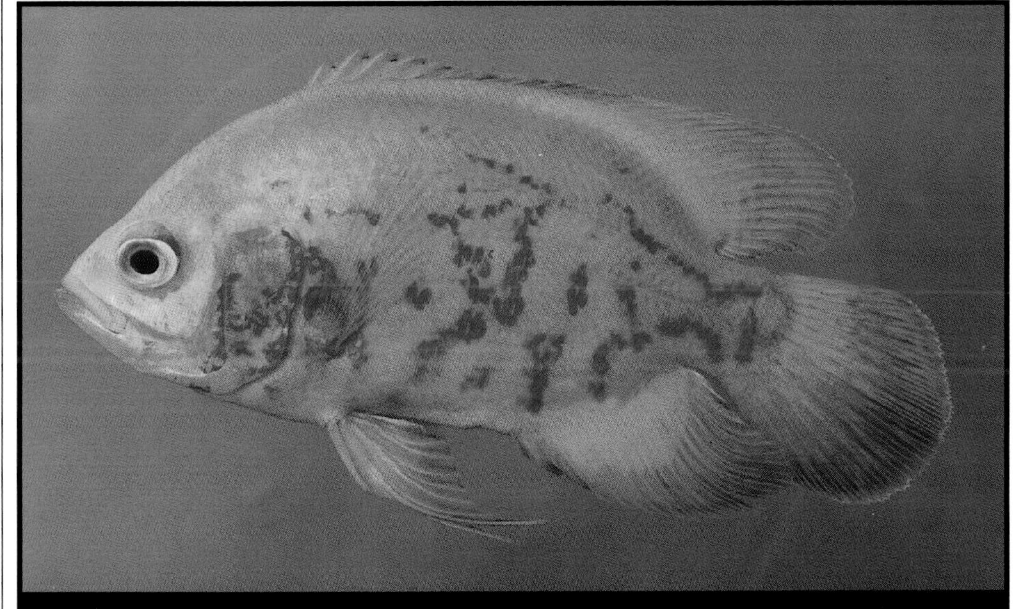

The lutino oscar looks like an albino but it has black pigmented eyes. Photo by Dr. Warren Burgess.

VARIETIES OF OSCARS

"IMPROVING" ON NATURE

This section should be subtitled "Many Varieties and a Cynic," as I tend to look with a jaundiced eye on fancy varieties of perfectly good fish species that are created by selective breeding. I have never failed to be amazed at how hobbyists will become enamored of a species because of the fact that they appreciate the animals as they exist in nature, but then they are diverted to attempts to create artificial varieties. To fully comprehend just how extreme my viewpoint is here, the reader needs to know that I think that the most beautiful angelfish is still *Pterophyllum scalare* as it is found in the wild. I think that the so-called wild forms are prettier and more majestic than the albino, veil tail, marble, black, and any of the other varieties. I prefer my animals to be unspoiled by the shaping of the hand of humankind through their efforts of selective breeding.

While admittedly a cynic about this, I would point out that I am not a total lunatic about this subject. I myself have been involved in selectively breeding a brighter firemouth cichlid and a redder red devil. However, that is not because these fish don't exist in nature. It is the case that only a small percentage of red devils are really bright red in their natural habitat (and that very fact has been the object of great study by scientists), and it makes sense

Some specimens of tank-raised oscars have an elongated shape and have two ocelli on their caudal peduncle. Fortunately, this variety has not found favor with aquarists and has disappeared. Photo by Wolfram Ch. Schrey.

Taiwanese breeders are producing this oscar which they call the *wild oscar.*

to have the strain of red devils that is bred in captivity consist of uniformly red red devils.

Obviously, I am a little extreme in my viewpoint, and I probably would have been among those who pooh-poohed the automobile and would have recommended that everyone stick to horses! And it certainly is not my intention to discourage oscar breeders from breeding some of the artificial strains or starting a new one of their own. I have purposely included this section at this particular point because I felt that an oscar breeder who has just recently succeeded in joining the fraternity of those who have bred oscars eventually has a decision to make. Should I breed regular oscars or a fancy variety?

Although fancy varieties initially bring a higher price, they always sink down to, or below, the price of "regular" oscars. If you are going to breed a fancy variety, your reasons should be that you really like the artificial strain. You may even want to improve on it, both in its hardiness and the intensity of the coloration. Whatever the situation, I thought that I would list the varieties that are presently extant. Doubtless there will be more by the time this book goes to press. I'll start with the original wild form.

THE WILD OSCAR

Although the wild-type oscar is still popular among aquarists and is widely available at aquarium shops, most of these are not wild at all. They have been bred for many generations on fish farms all over the world and in the tanks of dedicated hobbyists. This is an advantage in that the animals are free from disease, and the aquarium strain is certainly easier to spawn than wild individuals, and individuals of this strain have adapted themselves well to a wide variety of water parameters and are completely at home in the aquarium.

Although the domesticated oscar is somewhat variable in appearance, most of them don't quite have the richness of coloration that is possessed by the specimens in the wild—although the truth is that there is quite a bit of variation in the wild specimens. Collectors and ichthyologists report that they get individuals from the same water that have different numbers of eye spots (ocelli) on the dorsal, as well as many that have only one. The main constant is that there is nearly always an eye spot on the *caudal peduncle*. That is true in the aquarium specimens, too. The front is an emerald to dark green coloration, and this coloration extends back to the middle of the dorsal fin, but it is interrupted by irregular dark blotches and a pattern of red reticulations.

With the wide range of oscars, it is not unlikely that there are many subspecies, and some ichthyologists have speculated that the entire "oscar complex" may consist of a variety of species. In that case, is the aquarium strain a mixture of

Astronotus ocellatus from the Rio Negro near Barcelos. This is the fish which Dr. Axelrod raised from a one inch specimen. The inset shows the fish as a 2 inch specimen. Photo by Dr. Herbert R. Axelrod. Inset photo by Mark Smith.

Intermediate tank-raised specimens show dark marks along the base of the dorsal and pectoral fins but only the caudal peduncle sports a true ocellus. Tank-raised varieties are not entitled to scientific names. Photo by Dr. Herbert R. Axelrod.

different species? Probably not, but it is something to think about.

TIGER OSCARS

This is my favorite of the artificial varieties. That is probably because it is but little changed from the wild form. The "improvement" amounts to a selection for more red coloration in the individuals, while retaining all the other colors.

These artificial varieties are started when an individual specimen shows a qualitative difference from the other fish. The fish is bred, and the custom has been, with livebearers for example, to breed the progeny back to the original variant or "sport." However, since oscars take a long time to mature and tend to mate for life, this normal procedure is more difficult to follow with this species. It is probably for that reason that there are not more varieties of oscars.

This is a weird albino (note the red eye) oscar which has what looks like black pigment on the outer margins of its fins (which means it is not an albino)! This fish is produced in Taiwan.

The Taiwanese lutino oscar features unusual black and red edging to its unpaired fins but has a black eye.

The red oscar from Taiwan shows the ocellus on the caudal peduncle but it is a very nice-looking fish.

When oscars are frightened they lose most of their characteristic color and manifest *fright coloration.* Photo by Dr. Herbert R. Axelrod.

THE RED OSCAR

The red oscar was probably produced as a natural consequence of breeders very selectively pairing individuals that showed a lot of the red coloration. In other words, it is very likely that the red oscar was an outgrowth of selective breeding of the tiger oscar. Someone came up with a sport that was nearly all red on the side, and selective breeding eventually produced this strain. This is only speculation on my part, however, and some have opined that a cross between *Astronotus ocellatus* and *Astronotus orbiculatus* produced this color strain.

Certainly, the red oscar is pretty and it is colorful. If you find it more colorful than the wild strain, feel free to concentrate your breeding efforts with these specimens. There will still be plenty of us left that will stick to perpetuating the natural strain.

THE ALBINO OSCAR

These appeared in the hobby several years ago, and they attained a certain popularity, although it has lagged behind that

A wild specimen of *Astronotus orbiculatus.* This fish is often called the tiger oscar. Photo by Dr. Herbert R. Axelrod.

The American-bred strain of the albino oscar. Photo by Murray Wiener.

of the tiger and red oscar strains, not to mention the wild, unspoiled strain. If bred long enough and with enough consistency, albinos appear in nearly any species of fish. It is nearly always a simple recessive trait, so it is mere child's play to breed and perpetuate such a strain once the "sport" appears. The only problem is that albinos are occasionally infertile.

If you keep or breed albino oscars, be aware that they are more sensitive to light because of the lack of pigment. For that reason, you will want to provide subdued lighting in the tank.

OTHER VARIATIONS

One of the variations that has appeared with just about any fish species is the long-finned variety. They have always been unattractive to me in all species of fish, but they are especially so with the oscar. After all, part of the appeal of the oscar is the roundness of the body and of the fins. So why spoil it all with this grotesque long-finned foolishness? Well, there must be a reason, for there are long-finned forms available in all the color varieties, but, so far, they haven't been popular.

Wild-caught specimens of the tiger oscar, *Astronotus ocellatus*. Photo by Klaus Paysan.

FUN WITH YOUR OSCAR

OSCARS ARE REAL PETS

Although oscar mania can consist of breeding numerous pairs of oscars, possibly of several different varieties, it is worth making the point that it is possible to enjoy your oscar and even learn a lot about the species without becoming involved with all of that.

Some of the great oscar scholars have been those people who kept only one specimen at a time. However, the individual kept was as much a member of the family as any cat or dog. These may be the people who actually have the most fun with their fish. They only dedicate a fifty-or a hundred-gallon tank to one fish, but they raise it, train it, and devote lots of time to it. Not only may these people have the most fun with their fish, but the fish may particularly enjoy its time on earth too. Such an individual receives lots of special attention, as usually it is greeted first upon any home arrivals, and special time in the evening is spent with it, too.

I have personally known couples who had kept other fish before; in fact, most of them started out with a community tank. But they ended up keeping just one oscar as a pet. Not only did this individual take the place of any dog or cat, but it took the place of a child, too. A fish that has that position in the family truly receives bountiful attention.

Thus it is that oscars give more to humankind than just food. (Yes, the oscar is an important food fish in its natural habitat, but perhaps you should not let your oscar hear of this. Remember, they are sensitive!) Certain individuals are valued company and, perhaps, surrogate children for childless couples. This is not a contribution that is easily dismissed. And the good thing about oscars is that you don't have to pay for their college education, and they don't take the car out and wreck it.

Having known many people who owned an oscar for a pet, I have a few suggestions that may help enrich your life with a particular oscar that you may have picked to share your domicile.

TRAIN YOUR OSCAR TO DO TRICKS

As smart as oscars are, it must be remembered that they are still fish and that they lack a cerebral cortex. For that reason, we must keep the tricks simple. Even so, they will amaze people who don't think that fish can think. In attempting to train your fish, you need to have in mind the behavior you want and decide how you are going to encourage it.

The first step is to utilize a small bell or clicker. Ring the bell whenever you feed the fish. Eventually, your fish will come to associate the bell with the food. And it is easier to reward a behavior with a signal, such as a bell, than it is with the actual food itself.

For example, if you were going to teach your fish to jump through a hoop (and this is most definitely not recommended!), you would start by putting a hoop in the water. The fish would examine it, eventually sticking his head part way through. That is a small start, so you ring the bell and reward him with food whenever he does that. This goes on for several days, and the fish starts to get the idea that he can get food just by sticking his head in the hoop. But you begin to wait for him to move his head farther through the hoop before you reward him. Eventually, he has to swim through the hoop before he gets the reward. You get the idea. You gradually raise the hoop a tiny bit out of the water. Over a period of weeks, you eventually have a fish that will jump through a hoop. (I have actually seen an oscar that would do this, but the guy who owned him had a large fiberglass tank in his fishroom, and that was where the oscar was kept. There was no worry about water splashing on the floor, and the tank was covered with a screen grating, so there was no worry about the fish practicing on his own and possibly jumping out of the tank.

Some people like to have an oscar act like a watchdog, hitting the front glass and acting aggressive toward people. That is easily accomplished; in fact, some oscars demonstrate the behavior *without* the training. All it takes is a little encouragement for the oscar to demonstrate his natural propensity to protect his territory. Go up to the glass. Shake your head back and forth to mimic a challenging fish. And then back away as the oscar approaches. Do this a little while each day, and you will be surprised how soon you have a "watch oscar" on your premises. Only it is really difficult to get them to bark!

The trick that most people want is to have their pet fish actually allowing themselves to be petted. Although this is perhaps the most common "trick," it is among the least natural behavior for the oscar to learn. First, the hard facts: oscars have no natural desire to be petted. Dogs and cats are different in this respect. It is a natural behavior for them that has been enhanced by selective breeding over countless centuries. So, in the case of your oscar, you are going to need to provide lots of time and attention. Most people who pet their oscars have gradually gotten their fish used to being stroked. It may have started as something they did when they cleaned the tank, or the hobbyist may have deliberately set out to establish the behavior. What it amounts to is that the fish tolerates the touching and isn't spooked off by it because he has become used to it. That means lots of practice and starting things off very slowly. It also helps to reward the fish with food. In fact, sometimes the behavior is an outgrowth of hand-feeding your

Oscars can move their eyes. Note that their mouth is devoid of long, sharp or cutting teeth. Photo by Dr. Herbert R. Axelrod.

oscar. Teaching oscars to eat from your hand is as easy as falling off a log. Once you get them used to that, they begin to associate your hands with food and are therefore more willing to be touched by them. Just do it ever so gently at first. Make sure that you don't spook your fish, as it will be a long road back from that, as the fish will retreat whenever you put your hands in the water. Patience, as always, is the emphasis here.

PROVIDE YOUR OSCAR WITH TOYS

Aquarium shops display a number of decorations, from small boats to rocking chairs to small glass or plastic spheres for the aquarium. Taking into account an oscar's natural inclination to explore things and to mouth them, it is not surprising that oscars actually like having something new put into the tank each week. Just be sure to remove the old one and make sure that what you put in has no part that the oscar can ingest and possibly harm himself with.

The point is that oscars are highly curious, and they enjoy having something new put into the tank that they can toss around and examine. That doesn't mean that you have to buy something new each week. Just

The American-bred bronze oscar, often sold as the red oscar but it is a different strain.

get four or five items and rotate them through the tank weekly. Not only is it fun to watch the oscar playing with its new toy, but the habit of providing them to him may actually make him smarter. Studies with rats utilizing an enriched environment of new toys each week indicated that the brains of the rats which had experienced the enriched environment during their life-times actually had a larger brain mass as compared to the others. And oscars live longer than rats.

For some people, keeping an oscar is like keeping a family pet. They would no more think of breeding oscars than they would having a kennel of dogs. Even so, such persons may be excellent scholars of the oscar. But when watching these people with their pets, it is sometimes difficult to tell just who is studying whom!

For others, the real enjoyment of oscars is to keep them as a pair and be able to observe their family life.

Whichever type you are, you will find that oscars are intriguing fish indeed. It is not without reason that the very personable and affable oscar has maintained such prominence in the tropical fish hobby for so long.

INDEX

Page numbers in **boldface** refer to illustrations.